STEAM
in the coalfields

G T HEAVYSIDE

DAVID & CHARLES

NEWTON ABBOT · LONDON · NORTH POMFRET (VT) · VANCOUVER

ISBN 0 7153 7323 4

Library of Congress Catalog Card Number 76–54079

Set in 10 on 11 Times
and printed in Great Britain
by Redwood Burn Limited, Trowbridge & Esher
for David & Charles (Publishers) Limited
Brunel House Newton Abbot Devon

Published in the United States of America
by David & Charles Inc
North Pomfret Vermont 05053 USA

Published in Canada
by Douglas David & Charles Limited
1875 Welch Street North Vancouver BC

Contents

Introduction

SINCE the sixteenth century the coal industry has played a vital role in the industrial and economic growth of the United Kingdom. It is, however, true to say that many people very easily forget the great benefits that have accrued to mankind from excavating the mineral so kindly deposited by nature many millions of years ago, and when thinking of coal mining people tend to remember the difficulties which have beset the industry, the dangers which confront those who go down the shafts, and the despoilation of the country-side which has often been caused in times past.

But to many railway enthusiasts in the 1970s thoughts of collieries mean only one thing—the chance to see working steam locomotives in the environment for which they were originally intended, for since August 1968 when British Rail banished steam from its tracks on normal services, the National Coal Board has been the operator of the largest fleet of steam engines in the British Isles. It is appropriate that much of the commercial working life of the iron horse should be eked out in the coalfields, for this was the industry that gave birth to railways and fostered the early development of the steam locomotive.

The properties of coal have been known for centuries but its early use was limited because of the problem of carriage due to its great bulk and weight, and the only satisfactory way over any great distance was by river or sea. Thus the sale of coal from most pits was limited to local land sales, unless they were near navigable waters. Difficulties then arose as early reserves were worked out and later collieries were forced further inland. Huntingdon Beaumont eased the problem in 1604 by constructing a wooden waggonway of about two miles to carry coal from Strelley to Wollaton, near Nottingham, this reputedly being regarded as Britain's first railroad. Others copied the idea and during the next 200 years, after a slow beginning, many such waggonways were built, particularly along the banks of the Tyne and Wear, motive power being combinations of gravity, horses, men, and ropes.

Transport continued to be a major obstacle but progress was made during the 80 years from 1760, which was to become known as the Canal Age, when the rapid increase in the mileage of inland navigable waterways opened up new coal reserves and heralded the industrial revolution. Many miles of horse tramways were also constructed, both to act as canal feeder routes and to bridge gaps over difficult terrain. It was on one of these tramways from Pen-y-darren Ironworks, near Merthyr Tydfil, that Richard Trevithick, the inventor of the first successful steam locomotive, tried out one of his products in 1804, following his first trials with a locomotive at Coalbrookdale the year before, but it was left to men in the coal industry to make this revolutionary idea a viable commercial prospect.

It was not until 1812 that the first steam locomotives started regular service—on the Middleton Colliery Railway, Leeds, which in 1758 had been the first railway to be sanctioned by Act of Parliament. These locomotives built by Fenton, Murray and Wood in Holbeck, Leeds, to the patent of John Blenkinsop, the colliery agent at Middleton, were on the rack-rail principle, Blenkinsop not then having sufficient faith to rely on adhesion

The thoughts of an engine bearing the name of an abbey will fill the minds of many people with memories of the GWR Castle class, but, while they have long since departed from the main lines, *Llantanam Abbey*, in the shape of AB No 2074 built 1939, lives on at Mountain Ash, Glamorgan, here blowing off impatiently on a wet afternoon at the entrance to Deep Duffryn Colliery on 23 May 1973.

Threading the Doon Valley in the Ayrshire hills, between Minnivey Colliery and the tip, with coal for Dunaskin Washery from Pennyvenie Colliery on 30 August 1973, is NCB 0–4–0ST No 19 (AB No 1614), built 1918.

alone. Middleton returned to horse haulage in 1835 and only one or two other colliery lines were equipped with this system.

We turn to the North East for the next significant developments, where in the expanding Northumberland and Durham coalfields there was an increasing need to find a cheaper and more efficient way of moving the black diamond. A Trevithick-designed locomotive had been tried successfully in 1805 at Wylam Colliery, over the five mile waggonway to the Tyne, but was quickly relegated to a stationary boiler because of damage to the wooden track. This was replaced with an iron plateway in 1808, following which there were further experiments, and by 1813 locomotives were in use.

Others were also experimenting, including George Stephenson, destined to become known as 'the father of railways', and his first locomotive *Blutcher* began work in 1814 at Killingworth Colliery, where he was then employed. This was the first locomotive to run on edge rails and over the next ten years a number of his engines entered colliery service, the period culminating in the founding of the first locomotive works in 1823, Robert Stephenson & Co at Newcastle-upon-Tyne. Partners in the firm were George Stephenson and his equally famous son, Robert, and two others.

In the early 1820s thought was given to the construction of a railway, in preference to an earlier scheme for a canal, to connect the Aukland coalfield to the navigable Tees, and eventually this was opened in 1825 as the Stockton & Darlington Railway, the first public railway to use locomotives. While this was conceived as a parochial venture with the principal aim of transporting coal, little was it realised that it marked the start of a new era that was to alter radically the course of history. From the small beginnings of the S&DR grew the vast web of railways that ultimately linked practically every town in the United Kingdom, bringing travel possibilities undreamed of before the advent of this

To the delight of the enthusiasts No 7 of Littleton Colliery, Staffs, roars across the western outskirts of Cannock Chase, at the same time earning her keep by bringing a rake of empties back to the colliery. This engine is normally spare but was brought out on a colliery open day on 16 November 1974.

new age, and furthering the cause of industrial expansion by opening up more of the coalfields to the outside world.

Initial progress in track-laying was naturally slow, but from 1840 to 1870 the tentacles of iron rail extended at a phenomenal rate, with route mileage increasing from 1,500 to 15,500. Simultaneously, and indeed throughout the nineteenth century, the demand for coal grew greater and most of the new output was carried by this new and more efficient means, in addition to trade won from barge and shipowners, although they have never been totally eclipsed.

Locomotives were now required in ever-increasing numbers to handle the traffic, and in the mid-nineteenth century many factories were founded to meet this need, while some already established engineering firms tried their hand at locomotive construction. By the 1870s most railway companies had built their own workshops and the private locomotive industry had to look abroad, or to small industrial concerns such as collieries, in order to find a market for their products.

The first tank engines emerged in the 1840s, and their suitability for shunting and both-direction running without turning commended itself to the entrepreneurs of the coalfields, in the latter half of the nineteenth century, who purchased tank engines in ever increasing numbers. A few tender types, nevertheless, worked regularly over colliery metals. The duties were mainly yard shunting and short haul jobs to exchange sidings with the main line railways, and traditional work conveying coal to canal and river staithes. A number of the larger coal concerns, however, developed quite extensive systems running to many route miles, linking pits, washeries, coking plants, etc., some installing quite sophisticated signalling methods in keeping with those of their main line brothers.

Coal transport was not the sole use of the tracks, many colliery owners providing 'paddy'

8

Polkemmet Colliery, Whitburn, is famed for the double-heading of its trains up the steep incline from the colliery yard to the BR exchange sidings situated on the edge of Polkemmet Moor. A brace of grimy AB 0–6–0STs, 1175/1909 (leading) and 2358/1954, NCB Nos 8 and 25 respectively, approach the exchange sidings on 23 May 1974.

trains, often of a very spartan nature, for the convenience of employees, while a few ran public services, the most noteworthy being the South Shields, Marsden & Whitburn Colliery Railway whose timetabled services were continued by the NCB until November 1953.

The opening years of the twentieth century followed a similar pattern to the latter half of the previous one with more and more locomotives being acquired to move the ever-increasing output from the British coalfields. Not only were engines purchased new, but others came second-hand, including many main line discards, while a few of the large coal companies, like the early pioneers, built their own, some utilising parts supplied by established locomotive manufacturers.

The zenith of the industry was reached in 1913 when coal production totalled 287 million tons, of which 73 million was exported. World War I marked the end of the years of expansion, when the immediate effect was the loss of many valuable overseas markets. Then followed the very difficult and depressive 20-year period leading to World War II, during which, in common with other industries, the purchase of new motive power took a downward turn, resulting in many locomotive builders going out of business.

At the end of World War II coal was still the prime source of power in Britain and with the industry in a·run-down state, political rumblings advocating nationalisation were soon to the fore, an ideal mooted in some quarters for many years previously. Thus, in July 1946 the royal assent was given to the Coal Industry Nationalisation Act under which all coal mines, together with ancillary undertakings, including their railways, were vested in the National Coal Board from 1 January 1947.

Under state ownership came a not surprisingly diverse collection of over 1,500 steam locomotives, dispersed among 540 locations, and previously owned by 260 different coal companies, some themselves formed by earlier amalgamations. While a large proportion came from the factories of the more prolific builders, such as Andrew Barclay, Hudswell Clarke and Peckett, there were examples from manufacturers who had built very few; in all, 40 firms were represented, plus locomotives built by the coal owners, and those from the works of the railway companies, including many from the pre-grouping era. These, along with a handful of diesel locomotives and the overhead electrics of the Harton Coal Co Ltd (the oldest in use since 1908), formed the motive power of Britain's first nationalised industry.

For administrative purposes the coalfields were divided into 48 areas, which in many respects have followed individual policies regarding the locomotive fleet. The NCB adopted a standard blue livery but many areas chose not to apply it, leaving the colours of the previous owners or choosing their own scheme. Numbering was another hotchpotch, some areas having no coherent system, others listing in one consecutive run, or identifying the engines at each colliery separately. Subsequent area amalgamations and transfers of locomotives confused this issue still further, and there have been occasions when two engines bearing similar numbers have worked at the same colliery, giving rise at Shilbottle Colliery, Northumberland, to an engine being numbered 38½ for identification purposes! Personalisation by way of names has been common in some parts and non-existent in others. While maintenance and servicing has normally been carried out on site, central workshops were established for major repairs and overhauls, although firms such as Hunslet have received a little of this work.

At vesting day the motive power generally was in poor condition, due mainly to the years of neglect during the war, and the fact that many locomotives dated back to the nineteenth century. The situation was, however, eased by a class of engine whose birthright is owed to Britain's involvement in a battle-torn world.

In 1942 the Ministry of Supply decided that a simple, robust, but powerful shunting locomotive was needed to assist the war effort, the engine having to be capable of two years' intensive service under the worst possible maintenance conditions. Thus was born the now famous Hunslet 0–6–0ST Austerity class which was to see service in many parts of Europe, and to prove so successful that 484 were built, a total surpassed by only a few main line classes, with production continuing for over 20 years. With the end of hostilities the War Department had many locomotives for sale as surplus to requirements and in 1947 over 50 Austerities entered colliery service, following the footsteps of the few that had already gone to the aid of the coal owners. The class proved highly suitable for colliery work and over 250 have served the industry at some stage in their careers, including products from the works of all six war-time builders. The NCB bought 77 new (75 from Hunslet and two from Robert Stephenson & Hawthorns), with the remainder purchased second-hand or even third-hand, including some from BR where the locomotives operated as LNER Class J94.

During the first nine years of the NCB to 1955 an annual average of 30 new locomotives entered service, main suppliers being Hunslet, and Robert Stephenson & Hawthorns with sizeable contributions from the works of Andrew Barclay and Hudswell Clarke, along with over 30 vertical-boilered engines built by Sentinel. A few locomotives were ordered from Peckett, Bagnall and the Yorkshire Engine Co. Over 75 per cent. were saddle tanks, with the majority of the comparatively few side tanks supplied by Hudswell Clarke and the remainder by Andrew Barclay and Robert Stephenson & Hawthorns.

This pattern was not to continue and the second half of the 1950s brought developments which had far-reaching effects on the steam locomotive population. In 1955 the publication of the British Railways modernisation plan, which sounded the death knell of steam on the state railway system, seemed at the same time to point the way to other industries, including the NCB, to realise the advantages of internal combustion engines. Following the report, the British private locomotive builders obtained few orders for steam, and since 1955 less than 30 have been purchased new by the NCB, and diesel traction has progressively increased its hold on the traffic.

These years also witnessed a dramatic change in the fortunes of the coal industry, which from World War II until 1956 struggled continually to meet the demand for its product in an expanding energy market. But from then on, home consumers, with government encouragement, were to look more and more to other forms of power, principally oil, and from 1957 the NCB found itself having to fight for its markets.

At the same time, the gradual introduction by local authorities of provisions under the 1956 Clean Air Act did not help the coal trade, causing many coal-burners to change fuels and forcing locomotive users to look closely at their motive power. In an effort to bring steam locomotives within the scope of the Act, by overcoming the problem of black smoke, the Hunslet Engine Co Ltd developed an underfeed stoker and gas producer system, the first engine equipped being Hunslet Austerity No 2876 in 1961, which worked at Waterloo Main Colliery, Leeds. Over 50 NCB engines were modified, some by Hunslet and others in NCB workshops, and while the bulk were Yorkshire-based, the distinctive conical-shaped chimney of a converted Austerity has also been found at work in Lancashire, the Midlands, and the North East. A few locomotives were converted besides Austerities, but how successful the system has been is debatable since it has often fallen into disuse and normal methods of firing re-introduced. A mechanical stoker devised by Thomas Hill (Rotherham) Ltd was tried on a few Austerities, and at Walkden, near Manchester, coke burning—a reversion to the fuel of the pioneer steam engines of the first half of the nineteenth century—was employed on engines working in smoke control zones in attempts to combat the legislation.

One of the most significant developments in steam locomotive design in recent years has been the Giesl ejector, which improves the efficiency of the exhaust blast and its effect on boiler performance, patented by Dr Adolph Giesl-Gieslingen of Austria, an innovation widely adopted in some countries, although BR decided against equipping any more of its stock after experiments conducted with two of its engines. However, the NCB became the inventor's most valued British customer following tests carried out in 1959 with Hunslet Austerity No 2859 at Baddesley Colliery, Baxterley, Warwickshire, a location well known as the home of Britain's last working Beyer-Garratt, one of two supplied to the coal industry. During the trials No 2859 performed more efficiently than hitherto on the notorious Baddesley inclines, and 46 Giesl ejector sets were subsequently ordered, the last not until 1968. The majority were fitted to the Austerity class, but engines of other makers' designs, some of elderly status, also found themselves working with the characteristic oblong chimney which is part of the ejector.

Despite these modern developments the run-down of steam continued, and it was thus rather surprising that after a lull in the purchase of new engines since 1957, the turn of the decade found frames being laid in the erecting shop of the Leeds-based firm Hudswell, Clarke & Co Ltd for five 0-4-0STs for the Barnsley area, which emerged during 1960-1. Even this was not the finale for yet three more engines were to be built in Leeds, this time Austerity 0-6-0STs at Hunslet, the first in 1962 for Nailstone Colliery, Leicestershire, while the final two, built in 1964, found employment in South Yorkshire. The NCB's previous interest in the second-hand steam locomotive market was also now virtually at an end.

Decline in coal consumption in Britain continued, and during the period 1965-70 the NCB accelerated its colliery closure programme, reducing the number from 530 to under 300, thus making many steam locomotives redundant.

Other factors have also affected the destiny of steam power. In the latter part of the 1960s the diesel engine increased its hold on the traffic since British Rail had large numbers of surplus diesel shunters available for sale following its rationalisation programme. BR No D2518 was the first one purchased in 1967 for use at Hatfield Colliery, Stainforth, in the Doncaster area, and during the next four years a further 70 were bought from BR. Use of larger wagons by BR has reduced the need for locomotive work at some pit heads, as has the introduction of merry-go-round working in the last 10 years—in which permanently-coupled trains of hopper wagons, with automatic loading and discharge on the move at low speed, run continuously between pits and power stations hauled by BR locomotives throughout. NCB locomotives are thus not needed for these services. Moreover, in recent years road haulage has also played an increasing role in the transport of coal.

Internally, in an effort to improve efficiency, the NCB has cut down locomotive work by installation of conveyor belts, and use of dumper trucks, etc, and in some instances by connecting existing collieries together underground, whereby all the output comes to the surface at a central point.

Among the railway enthusiast fraternity only a few took more than a superficial interest in the industrial steam locomotive until the late 1960s when the passing of steam from BR changed all this, and today places like Walkden, Waterside, and Whitehaven, are revered by many as affectionately as the great railway meccas of old, such as Dainton, Derby and Doncaster. I, in common with others, now regret that this interest lay dormant for so long, and while there were over 650 steam engines still on the books of the NCB in mid-1968, their reign was rapidly entering its final phase, for by January 1970 the total had been reduced to just over 450. While 10 engines of Victorian origin lived to see the 1970s and the stock was from the workshops of 20 builders, a large proportion of the work was monopolised by 150 Hunslet Austerity class locomotives, with valuable support from Andrew Barclay, Hudswell Clarke, Peckett and modern Robert Stephenson & Hawthorns locomotives. Nine former BR locomotives (excluding Austerities), comprising eight pannier tanks from the Western Region and an LMS Jinty Class 3F 0–6–0T, were alive in the coalfields at the dawn of this era.

During the 1970s the run-down of the stud has continued and many locomotives have given little by way of revenue-earning service, some spending years waiting for the breaker's hammer or acting as spare engine in case of need. But in numerous places locomotives, liveried in various shades of green, maroon, blue, or just plain black, have continued to give sterling service, often on track which can hardly be classed as up to main line standards, and sometimes scaling banks which make Shap and Hemerdon look flat by comparison. Engines have frequently not been in the best of health, but characteristically they have forged on when many lesser mortals would have given up long ago. Despite the general picture of gloom, on my travels round the coalfields, it has been a pleasure on many occasions to meet men who take a real pride in their charges, and who can fairly be compared with, and described as, the Sammy Gingell's and Ted Hailstone's of the NCB world.

At the time of going to press the NCB still owns over 120 steam engines, nearly half being Austerities, and about 25 per cent Andrew Barclay. Only 6 side tanks and 1 pannier tank remain. The engines are dispersed over 60 locations, but many are just rotting hulks awaiting the visit of the scrap-metal merchant, and on an average working day only about 20 are steamed, some for the morning shift only. But steam action can on occasions be seen in unexpected places as locomotives stand in for temperamental diesels, sometimes for quite long periods, and the recent energy crisis may yet prove a blessing in prolonging the life of steam, even if only for months rather than years.

Inevitably there will come a day when the steam engine will be seen no more in the setting of its early development, another casualty of the march of progress, but happily forces are at work to try and ensure that its memory is not confined to books and photographs. Already steam passenger services are a regular feature over former colliery lines at Foxfield, Staffordshire, and Middleton, Leeds, and there are schemes in the pipeline for the restoration of others.

Perhaps the most ambitious scheme in the field of industrial preservation is at Beamish, Co Durham, where the North of England Open Air Museum has been actively working on the reconstruction of a typical nineteenth-century colliery, complete with railway, rope-worked incline, staithes, etc, as part of a large project depicting many aspects of life in the North East in the last century. Locomotives have already been obtained, including some from the NCB.

In recent times the NCB has been very generous towards railway preservation and in addition to the two engines at the Lound Hall Mining Museum, near Retford, many locomotives have been made available to both static and working railway museums, including examples from the pre-grouping years such as a North Staffordshire Railway 0–6–2T and Mersey Railway 0–6–4T *Cecil Raikes*, and which, but for the transfer of their allegiance to the coal industry, would have been lost to posterity years ago. Locomotives have been purchased from the NCB for use on preserved lines, while some ex-BR preserved locomotives have found homes at NCB premises, the most famous being A4 Pacific *Sir Nigel Gresley* at Philadelphia, Co Durham, and sister engine *Bittern*, along with A2 Pacific *Blue Peter* at Walton Colliery, Crofton, near Wakefield. The NCB's benevolent attitude

The driver of *Gamma*, at Vane Tempest Colliery, Seaham, checks the water level as BR Class 37 No 37045 waits to leave the colliery yard on 28 August 1974. Another photograph of this engine appears on page 38.

has also enabled enthusiasts to savour the delights of steam in surroundings varying from the stark industrial to the strictly rural, including trips in special trains over some of the more extensive systems.

In this book I have endeavoured to portray the steam engines in the British coalfields as I have known them during the 1970s, with photographs arranged in the form of a nationwide tour, grouped under NCB areas as formed in March 1967, except that for convenience the two Scottish areas have been regarded as one unit. Many of these areas now no longer exist as separate entities following further amalgamations. Included are photographs taken at coking plants and while these locations now come under the auspices of National Smokeless Fuels Ltd—a wholly-owned subsidiary of the NCB—they are featured under the relevant geographical area, as are preservation sites, etc. The maps of five of the systems show the basic layouts at the time of my visits, and today some of these look very different in this world of change. Engines have been identified by their builders' works number and year of manufacture. In the text no account has been taken of county boundary changes under the 1974 re-organisation.

My only regret is that I was not able to work in this field at an earlier stage, for much of interest has escaped, but I am thankful that I have been in time to record on film at least a part of the scene. I shall be very sorry to see the final demise of the iron horse from the industry it has served so well, but when that dreaded day dawns I hope that this book will form a suitable epitaph to a faithful servant, that it will revive happy memories for those who knew this age, and that some of its character and atmosphere will be conveyed to those who knew it not.

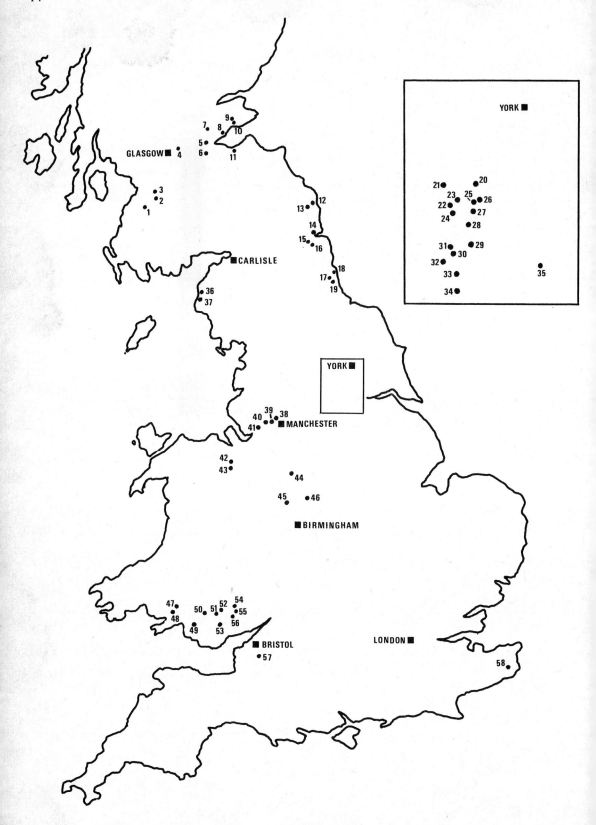

14

KEY TO MAP

SCOTTISH AREAS (NORTH AND SOUTH)

1 Waterside system, Ayrshire—serving Minnivey colliery, Pennyvenie colliery and Dunaskin washery.
2 Barony colliery, Auchinleck, Ayrshire.
3 Mauchline coal preparation plant, Mauchline, Ayrshire.
4 Bedlay colliery, Glenboig, Lanarkshire.
5 Kinneil colliery, Bo'ness, West Lothian.
6 Polkemmet colliery, Whitburn, West Lothian.
7 Comrie colliery, Saline, Fifeshire.
8 Pittencrieff Park, Dunfermline, Fifeshire.
9 Thos Muir's scrap-yard, Easter Balbeggie, Fifeshire.
10 Frances colliery, Dysart, Fifeshire.
11 Newcraighall Landsale yard, Niddrie, Midlothian.

NORTHUMBERLAND AREA

12 Shilbottle colliery, Shilbottle.
13 Whittle colliery, Newton-on-the-Moor.
14 Backworth system—serving Eccles colliery, and Fenwick colliery.

NORTH DURHAM AREA

15 Norwood coking plant, Gateshead.
16 Ravensworth 'Park Drift' mine, Lamesley.

SOUTH DURHAM AREA

17 South Hetton system—serving South Hetton colliery, and Hawthorn colliery and coking plant.
18 Vane Tempest colliery, Seaham.
19 Shotton colliery, Shotton.

NORTH YORKSHIRE AREA

20 Peckfield colliery, Micklefield.
21 Middleton railway, Leeds.
22 Newmarket Silkstone colliery, Stanley.
23 Savile colliery, Methley.
24 St John's colliery, Normanton.
25 Wheldale colliery, Castleford.
26 Fryston colliery, Fryston.
27 Glasshoughton coking plant, Castleford.
28 Ackton Hall colliery, Featherstone.

BARNSLEY AREA

29 South Kirkby colliery, South Kirkby.
30 North Gawber colliery, Mapplewell.
31 Woolley colliery, Darton.
32 Dodworth colliery, Dodworth.
33 Skiers Spring drift mine, Wentworth.
34 Smithy Wood coking plant, Chapeltown.

DONCASTER AREA

35 Markham Main colliery, Armthorpe.

NORTH WESTERN AREA

36 Harrington coal preparation plant, Lowca, Cumberland.
37 Whitehaven system, Cumberland—serving Haig colliery, Ladysmith washery, and Whitehaven harbour.
38 General Engineering workshops, Walkden, Lancashire.
39 Astley Green colliery, Tyldesley, Lancashire.
40 Bickershaw colliery, Leigh, Lancashire.
41 Bold colliery, St Helens, Lancashire.

42 Gresford colliery, Wrexham, Denbighshire.
43 Bersham colliery, Rhostyllen, Denbighshire.

STAFFORDSHIRE AREA

44 Foxfield Light Railway, Dilhorne.
45 Littleton colliery, Huntington.

SOUTH MIDLANDS AREA

46 Cadley Hill colliery, Swadlincote, Derbyshire.

WEST WALES AREA

47 Graig Merthyr colliery, Pontardulais, Glamorganshire.
48 Brynlliw colliery, Grovesend, Glamorganshire.
49 Maesteg system, Glamorganshire—serving Caerau colliery, Coegnant colliery, St John's colliery, and Maesteg washery.

EAST WALES AREA

50 Mardy colliery, Maerdy, Glamorganshire.
51 Mountain Ash system, Glamorganshire—serving Deep Duffryn colliery, Mountain Ash; Penrikyber colliery, Penrhiwceiber; and Aberaman phurnacite plant, Abercwmboi.
52 Merthyr Vale colliery, Aberfan, Glamorganshire.
53 Tymawr colliery, Pontypridd, Glamorganshire.
54 Blaenavon coal preparation plant, Blaenavon, Monmouthshire.
55 Talywain landsale yard, Abersychan, Monmouthshire.
56 Hafodyrynys colliery, Pontypool, Monmouthshire.
57 Kilmersdon colliery, Radstock, Somerset.

KENT COALFIELD

58 Snowdown colliery, Nonington.

LOCOMOTIVE BUILDERS

Locomotives featured have been built by the following firms and the given abbreviations used throughout.

AB Andrew Barclay, Sons & Co Ltd, Caledonia Works, Kilmarnock.
AE Avonside Engine Co Ltd, Fishponds Works, Bristol.
EV Ebbw Vale Steel, Iron & Coal Co Ltd, Ebbw Vale.
GH Gibb & Hogg Ltd, Airdrie.
HC Hudswell, Clarke & Co Ltd, Railway Foundry, Leeds.
HE Hunslet Engine Co Ltd, Hunslet Engine Works, Leeds.
HL R & W Hawthorn, Leslie & Co Ltd, Forth Bank Works, Newcastle-upon-Tyne.
MR Motor Rail Ltd, Simplex Works, Bedford.
MW Manning, Wardle & Co Ltd, Boyne Engine Works, Leeds.
NB North British Locomotive Co Ltd, Hyde Park, Queen's Park, and Atlas Works, Glasgow.
P Peckett & Sons Ltd, Atlas Locomotive Works, Bristol.
RSH Robert Stephenson & Hawthorns Ltd, Darlington and Newcastle-upon-Tyne.
WB W G Bagnall Ltd, Castle Engine Works, Stafford.
YE Yorkshire Engine Co Ltd, Meadow Hall Works, Sheffield.

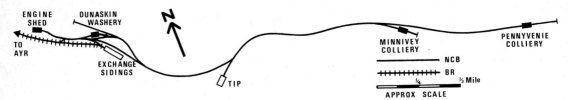

Above: Diagram of the NCB Waterside system situated high in the Ayrshire hills by Dalmellington.

SCOTTISH (NORTH AND SOUTH)

Below: Sandwiched between side-tipping wagons, after returning from the tip, AB No 2335 of 1953, NCB 0–6–0T No 24, waits to push one set into Minnivey Colliery yard before proceeding with the remainder to Pennyvenie on 21 May 1974. The Giesl ejector was fitted in 1965.

Above: Later on the same day No 24 is about to leave Pennyvenie Colliery on the 3½-mile journey to the washery with the morning's output.

Below: Shunting at Dunaskin Washery, Waterside, on 27 May 1974 NCB 0–4–0ST No 10 (AB 2244 of 1947) makes a volcanic effort in moving a string of NCB internal wagons.

Above: Built in 1913 the veteran of the Dunaskin Barclay fleet No 1338, NCB 0–6–0T No 17, assists No 10 at the start of the climb up the ferocious gradient to the back of the washery with coal which the latter had brought from Pennyvenie on 26 August 1974. Seen clearly are the wagons which act as tenders to supplement the engines' limited bunker capacity, this also being a feature at some other Scottish industrial locations.

Right: Inside Dunaskin shed at the end of the morning shift the drivers and shunters prepare to leave for home, while No 24 simmers gently alongside No 17 on 21 May 1974.

AB 0–4–0ST No 1116, built 1910, in various moods as NCB No 16.

Above: In this view from the footplate at Mauchline Coal Preparation Plant, between Cumnock and Kilmarnock, on 29 August 1973, No 16 is in a passive mood, the fire having been dropped on completion of the day's work. The colliery at this site closed in 1966, but coal from other local pits continued to be washed here until January 1974.

Right: At Barony Colliery, Auchinleck, Nr Cumnock, on 26 August 1974, a few days after being transferred from Mauchline, No 16, although in a fighting mood, fails to surmount the short incline to the BR exchange sidings and is seen stalling on the last of its five attempts.

Left: Ignominy as AB DM No 417 of 1957 comes to the rescue of the ailing veteran.

Top left: BR Class 20 No 8080 prepares to leave Kinneil Colliery, Bo'ness, with a trip working to Bo'ness Junction on the Edinburgh–Glasgow main line, as the colliery shunter, NCB 0–4–0ST No 21 (AB 2292 of 1951), awaits its next duty on 12 May 1972.

Above: On the same day Glenboig cattle graze peacefully near the colliery tip, ignoring the efforts of No 17 hauling coal to the exchange sidings.

Left: Within sight of the tenements of Glasgow yet set amidst countryside is Bedlay Colliery, Glenboig, where on 22 May 1974 the colliery 'pug', by which term Scottish industrial locomotives are affectionately known, pulls wagons away from the screens. The engine is another AB 0–4–0ST, works No 2296 of 1952, NCB No 17.

Above: On 10 May 1972 Victorian Barclay, 0–6–0ST, No 885 of 1900—a second NCB No 8 at Polkemmet—pilots HE Austerity 0–6–0ST No 2880 of 1943 NCB No 17, away from the colliery.

Below: The driver of No 25 at Polkemmet Colliery, Whitburn, cleans out the smokebox. This engine is portrayed in action on page 9.

Above: In recent times the work at Comrie Colliery, Saline, Fife, has been in the hands of Austerity 0–6–0STs, where on 22 May 1974 NCB No 19 (HE 3818/1954), is shunting wagons over the weighbridge. On the left is the Scottish Rexco Ltd smokeless fuel plant for whom traffic is also carried over the branch from the exchange sidings at Oakley.

Top right: Giesl ejector-fitted Austerity 0–6–0ST No 7 (WB 2777 of 1945) captivates its young audience as it pushes empties for Comrie across the A907 Dunfermline to Alloa Road at Oakley on 11 May 1972. The signalbox on the left, operated by the shunter, controls the gates, signals, and catch-points, which protect the crossing.

Right: Following withdrawal at Cardowan Colliery, Stepps, this GH0–4–0ST, built 1898, was presented by the NCB to the Carnegie Dunfermline Trust in May 1968, and is now on display in Pittencrieff Park, Dunfermline. The gift is commemorated by a tablet which can be seen between the fence and the rear wheel. The photograph was taken on 24 May 1974.

Below: Engines located at Niddrie loco shed in the Edinburgh suburbs had little work following the closure of Newcraighall Colliery in 1968, that which remained being for a landsale yard which also closed in late 1972. On 8 May 1972 No 25, AB 2358 of 1954, takes water outside the shed, the engine looking a little cleaner than in its subsequent days at Polkemmet, to where it was transferred following the abandonment of operations at Niddrie.

Above: 1949 built AB 0–4–0ST No 2262 which ended its working life as No 7 at Frances Colliery, Dysart, languishing with other NCB and Wemyss Private Railway locomotives in Thos Muir's scrapyard at Easter Balbeggie, Nr Thornton, Fife, on 24 May 1974.

Right: On the same day at Frances Colliery overlooking the Firth of Forth, sister engine No 30, AB 2259/1949, proves she is still alive with this fine exhaust as she shunts full wagons in the exchange sidings.

Above: Set in beautiful Northumberland countryside just south of Alnwick, is Shilbottle Colliery, where on 31 August 1972 NCB No 45 (Austerity RSH No 7113, built in 1943) is drawing wagons out of the colliery yard. No 45 was purchased by the NCB from the Port of London Authority in 1960.

Top right: At nearby Whittle Colliery, Newton-on-the-Moor, on the same day, two more RSH products stand in the engine holding siding, No 47 0–6–0ST, 7849/1955, is waiting to start work while the commercial working life of No 31 0–6–0T, 7609/1950, is already at an end. On the right BR Class 08 No D3215, on hire to the NCB, is leaving the colliery on the 4½-mile journey to the exchange sidings on the East Coast main line. Both steam locos have since been moved to the North Yorkshire Moors Railway.

NORTHUMBERLAND

Right: No 47 darkens the sky as it leaves the holding siding.

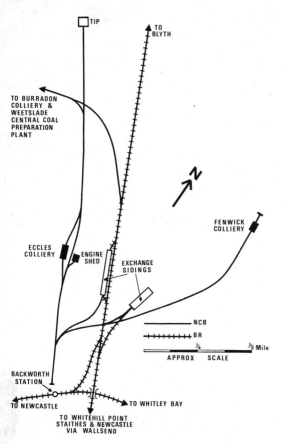

TIP

TO
BLYTH

TO BURRADON
COLLIERY &
WEETSLADE
CENTRAL COAL
PREPARATION
PLANT

N

FENWICK
COLLIERY

ECCLES
COLLIERY

ENGINE
SHED

EXCHANGE
SIDINGS

NCB

BR

¼   ½ Mile
APPROX   SCALE

BACKWORTH
STATION

TO NEWCASTLE

TO WHITLEY BAY

TO WHITEHILL POINT
STAITHES & NEWCASTLE
VIA WALLSEND

Left: Diagram of railways in the Backworth area, just north of the Tyne.

Right: Under the watchful eye of the flagman, NCB 0–6–0ST No 6, one of the Austerities, WB 2749 built 1944, which have been the mainstay of Backworth motive power in recent years, hauls fulls across the minor road at the north end of Eccles Colliery on 28 August 1974. The load will be taken past the colliery to the weighbridge at the Tyne end.

Below: No 9 RSH No 7097 of 1943 shows little respect for the Clean Air Act as she completely obliterates the view of the pit-head buildings at Fenwick Colliery on 1 September 1972. This colliery closed in August 1973.

Above: On its 3½-mile journey from Eccles Colliery to Weetslade Central Coal Preparation Plant, No 48 enters the loop after crossing the main street at Burradon village on 28 August 1974. Coal for washing had to be taken to Weetslade following the breakdown of the washer at Eccles.

Below: No 6 leaves the exchange sidings with empties for Eccles Colliery on 27 August 1974. On the right is the former NER Blyth and Tyne line which was crossed on the flat by the NCB line to Fenwick.

Above: No 9 shortly after leaving Fenwick on 1 September 1972.

Below: No 48, 2864 of 1943, one of the first batch of Austerities to emerge from Hunslet Works, pulls loaded BR hopper wagons out of Eccles Colliery, while No 6 waits to assist in moving the load to the weighbridge (see page 32) on 28 August 1974.

NORTH DURHAM

Top left: Resting in the shed at Norwood Coking Plant, Gateshead, on 28 August 1974 is RSH 0–6–0ST No 7412 of 1948 in company with HE DH No 6677 of 1967, the steam engine being retained as spare to the diesels.

Left: A little further south from Norwood and adjacent to the BR Tyne marshalling yard at Lamesley, NCB No 32 propels coal from Ravensworth 'Ann' Colliery, across a viaduct spanning the Team Valley trading estate on 30 August 1972. The coal is brought to the surface at the site of the closed Ravensworth 'Park Drift' mine. The railway closed in April 1973 but the engine, 0–4–0ST No 1659, built by AB in 1920, has been preserved by the Stephenson & Hawthorns Locomotive Trust at the Tanfield Railway Company headquarters at nearby Marley Hill.

SOUTH DURHAM

Above: Earlier on the same day, NCB Austerity No 62, HE 3687 of 1949, runs light from the modern Hawthorn Colliery and Coking Plant complex at South Hetton, as sister engine No 69 HE 3785 of 1953, fitted with Hunslet underfeed stoker, waits for the road with empties from Cold Hesledon, which is at the head of a rope-worked incline from Seaham Harbour. The junction, which also gives access to South Hetton Colliery, is controlled, along with the colour-light signals, from a signalbox just out of the picture on the left.

Left: Deputising for a diesel under repair at Vane Tempest Colliery, Seaham, on 28 August 1974 is No 2502/7 *Gamma*, WB Austerity No 2779 of 1945.

Above: No 62 at the entrance to South Hetton Colliery with empties from Seaham Harbour on 30 August 1972.

Above and below: On the same day, two years short of its three score years and ten, another 0–6–0ST AB 1015 of 1904 carries out some of the final duties at Shotton. Fortunately this septuagenarian has been saved from the graveyard by the Stephenson & Hawthorns Locomotive Trust.

Left: On 29 August 1972 at Shotton Colliery, Shotton, the notice board proudly proclaims that the shaft was sunk in 1833, but four days later coal production ceased. Already out of use lying under the shadow of the winding gear is P 0–6–0ST No 1310 of 1914.

Above: On the historic Middleton Railway, weekend steam passenger services are operated by the Middleton Railway Trust, and on Sunday, 21 July 1974, ex Associated Portland Cement Manufacturers Ltd, Swanscombe, HL 0–4–0ST No 3860, built 1935, nears Middleton Park Gates terminus, on part of the original 1758 route, with a train from Tunstall Road Halt. The proximity of the centre of Leeds can be gauged from the buildings in the left background.

Top right: At Peckfield Colliery, Micklefield, on 24 August 1972 No S121 *Primrose No 2*, HE 0–6–0ST No 3715 of 1952 approaches the tip with colliery waste, much of the locomotive work here being on these duties. In the right background is Micklefield station on the former NER Leeds to York line. The engine is now on the Yorkshire Dales Railway, Embsay, Nr Skipton.

NORTH YORKSHIRE

Right: An immaculate looking HC 0–6–0T (1822/1949), NCB No S100 begins the climb to the tip at Peckfield on 6 June 1972. This engine is also now on the Yorkshire Dales Railway.

Above: At Newmarket Silkstone Colliery, Stanley, Nr Wakefield, No S103 HC side-tank No 1864, built 1952, pushes empties towards the back of the screens on 23 August 1973.

Left: NCB No S111 *Airedale No 2* leaves a smokescreen while shunting at Savile Colliery, Methley, Nr Castleford, on 28 March 1974. The engine is HE 1956 of 1939.

Above: On her then weekly outing while a diesel locomotive was serviced, 0–6–0T *Cathryn* NCB No S102, HC 1884 of 1955, comes up-grade past St John's Colliery, Normanton, on 24 August 1972, with wagons from Park Hill Colliery, Wakefield, which had been brought most of the way by their diesel. Coal production ceased at St John's in July 1973, but the coal preparation plant remained in use.

Top right: Austerity HE 2879 built 1943 returns to the colliery yard at Wheldale, Castleford, on 6 June 1972, with empties from the Water Basin sidings, where coal is loaded into trains of water-borne containers known locally as 'Tom Puddin's', and then hauled by tugs along the Aire and Calder Navigation.

Right: On the same day at nearby Fryston Colliery, another of the HC 0–6–0 side tanks purchased by the NCB, 1883 of 1955, *Fryston No 2* heads some motley-looking wagons past *Rose Louise* HC DM No D972 of 1956 standing on the shed road.

Right: A beautifully turned-out HL 0–6–0ST No 3575 of 1923 *Coal Products No 3* shunts vigorously at Glasshoughton Coking Plant, Castleford, on 21 August 1974.

Below: Reflections of a forsaken Austerity at Glasshoughton. *Coal Products No 7* waiting for the cutters torch on 28 March 1974. The engine was built by Hunslet in 1943 as No 2897, and rebuilt by the same company in 1963 when it was renumbered 3886.

Bottom right: On the same day *Coal Products No 3* lifts more coal for processing in the Glasshoughton ovens out of the exchange sidings.

Left: Standing idle in the fitting shop at Ackton Hall Colliery, Featherstone, where the steam locos have seen little use in recent years, is 0–6–0ST No S119 *Beatrice*, HE 2705 of 1945, on 21 August 1974.

BARNSLEY

Right: Outside the shed at South Kirkby Colliery, South Kirkby, on 11 May 1971 is *York No 1*, YE 0–4–0ST No 2474, built 1949. Locomotive work here had been reduced considerably by this time following the installation of an overhead automatic loading bunker in connection with the working of merry-go-round trains by BR.

Above: The remains of 1945 built Austerity HE No 3208 at Woolley Colliery, Darton, on 28 March 1974, the boiler having been removed to Cadley Hill Colliery, Swadlincote, as a spare for their Austerity.

Left: Monckton No 1, HE Austerity No 3788 of 1953, viewed through the shattered shed windows at North Gawber Colliery, Mapplewell, on 28 March 1974.

Top right: Private dwellings surround the colliery at North Gawber, an ample reminder of the need for effective smoke-reducing equipment. Shunting the colliery on 11 May 1971 is Austerity HE 3212 of 1945 fitted with Hunslet underfeed stoker unit and also a headlight.

Right: Coasting gently downhill at Dodworth Colliery, Dodworth, on 11 May 1971 is 1960 built 0–4–0ST HC No 1889. To the right of the pylon, in the distance, can be seen a BR class 37 on the former GCR Penistone to Barnsley route.

Above: Two of the final batch of five steam locos built by Hudswell Clarke can be seen in this view at Skiers Spring drift mine, Wentworth, on 22 August 1972, which forms part of the Rockingham Colliery complex. Both were delivered in 1961, No 1891 is transferring a couple of wagons to the exchange sidings, while on the right, out of use, is No 1892.

Right: At Smithy Wood Coking Plant, Chapeltown, just north of Sheffield, on 24 June 1974, Austerity *SWCP No 1* moves away from the quencher, having been summoned there following a temporary breakdown to the electric locomotive on coking car duties. The engine is another Hunslet rebuild, No 3887 of 1964, originally No 3193 of 1944.

DONCASTER

Above: Few steam locomotives have been seen in action in the Doncaster area in recent years, but on 29 March 1974 0–6–0ST 1953 built HE No 3782, makes a fine sight at Markham Main Colliery, Armthorpe.

NORTH WESTERN

Right: NCB crest as carried by many locos in the North Western Area.

Below: Coal winding stopped at Harrington No 10 Colliery, Lowca, north of Whitehaven, in July 1968, but the washing plant continued in use for the output from Solway Colliery, Workington, until May 1973 when both sites closed. On 14 May 1971 Giesl ejector-fitted Austerity *Warspite*, HE 3778 of 1952, attacks the last few yards of the climb to the plant with coal collected from the exchange sidings.

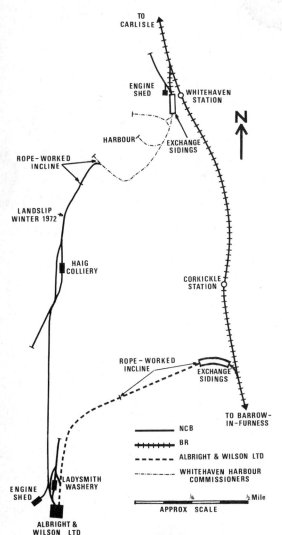

TO
CARLISLE

ENGINE
SHED

WHITEHAVEN
STATION

N

HARBOUR

EXCHANGE
SIDINGS

ROPE-WORKED
INCLINE

LANDSLIP
WINTER 1972

HAIG
COLLIERY

CORKICKLE
STATION

ROPE-WORKED
INCLINE

EXCHANGE
SIDINGS

TO BARROW-
IN-FURNESS

	NCB
	BR
	ALBRIGHT & WILSON LTD
	WHITEHAVEN HARBOUR COMMISSIONERS

APPROX SCALE

½ Mile

ENGINE
SHED

LADYSMITH
WASHERY

ALBRIGHT &
WILSON LTD

Left: Diagram of railways at Whitehaven. Before a land-slip during the winter of 1972 coal from Haig Colliery, after making the return journey along the cliff top to Ladysmith Washery, was despatched via the rope-worked incline to the harbour for shipment or transfer to the BR exchange. Severance of the direct route to the harbour means that all coal must now leave Ladysmith by road or over Albright & Wilson Ltd tracks.

Right: This panoramic view of Whitehaven was taken on 14 May 1971 as 0–4–0ST No 8 (AB No 1974 of 1931) wheels coal round the harbour from the foot of the rope-worked incline.

Bottom right: No 8 skirts the harbour with coal from Haig Colliery, and is about to pass Whitehaven Harbour Commissioners 0–4–0ST *Victoria* (P No 2028 of 1942), an engine often hired to the NCB, on 13 May 1971.

Below: Repulse, Austerity HE 3698 of 1950 with Giesl ejector at Ladysmith on 28 August 1973.

Above: AB 1448, built 1919, *King*, one of the few 0–4–0STs fitted with a Giesl ejector, pauses in Haig Colliery yard on 30 August 1974.

Right: On the same day another Giesl-fitted Austerity *Respite*, HE 3696 of 1950, begins the assault on the climb to Ladysmith, away from what is now the end of the line at Haig. Behind the engine can be seen a wagon on the isolated track beyond the site of the landslip and also the Irish Sea from under which the Whitehaven coal is wrought.

Left: Refreshment for *Repulse* at Ladysmith, on 28 August 1973.

Left: On 20 August 1970 at Astley Green Colliery, Tyldesley, Lancs, Austerity HC 1776 of 1944 *Harry* disturbs the peace of the fisherman (is he also a steam enthusiast?) on the Bridgewater Canal. Winding at this the last pit on the well-known and extensive Walkden system had ceased the previous April but coal was washed here for a further six months, when locomotive work came to an end.

Bottom left: On a murky morning in October 1970 a Giesl-fitted Austerity has a final flurry at Astley Green while engaged on clearing-up duties.

Below: Good omen for the future? *Respite*, HE 3696 of 1950, stripped down in the NCB General Engineering Workshops, Walkden, Nr Manchester, on 23 January 1975 following an accident at Haig Colliery, Whitehaven, during November 1974.

Above: Austerity *Hurricane* lives up to its name while storming away from Bickershaw Colliery, Leigh, on 4 August 1972. The locomotive was purchased new from Hunslet in 1955 as works No 3830.

Top right: At Bold Colliery, St Helens, yet another Lancashire-based Austerity *Whiston*, HE 3694 of 1950, takes a short sojourn by the weighbridge as excess coal is removed from an overloaded BR automatic discharge wagon on 20 August 1974. From the weighbridge the wagons have to be taken up a gradient so steep that even the powerful Austerities are limited to two full wagons on each journey.

Right: On the same day *Whiston* has its bunker replenished by a somewhat sophisticated method.

Above: In pristine condition 0–4–0ST *Shakespeare* of 1914 vintage, HL No 3072, bustles about its business at Bersham Colliery, Rhostyllen, Nr Wrexham, on 21 August 1973.

Right: Gwyneth RSH built Austerity, 7135 of 1944, runs light from the exchange sidings back to Gresford Colliery, Wrexham, on 5 June 1972. The scene is now but a memory, the colliery having closed in October 1973.

Above: Also on 5 June 1972 at Bersham, P 0–4–0ST No 1935 of 1937 *Hornet* is engaged at the side of the ex-GWR Chester to Shrewsbury line.

Foxfield Colliery, Dilhorne, near Stoke-on-Trent, closed in 1965 and the $3\frac{1}{2}$-mile line to Blythe Bridge on the BR Derby–Crewe route is now utilised by the Foxfield Light Railway Society, who run Sunday afternoon passenger services during the summer months.

Top right: The unique WB 0–6–0ST No 2193, built 1922, *Topham*, which gave nearly 50 years colliery service before being privately preserved, takes water at Foxfield ready for a working to Blythe Bridge on 28 July 1974. Nestling in the shed is 0–4–0ST *Hawarden* WB 2623 of 1940 which was acquired from the British Steel Corporation, Etruria, Stoke, while the colliery winding gear serves to remind visitors of the line's original purpose.

Right: This view from the footplate of *Topham*, as it begins the return journey from Blythe Bridge, was taken on the same day.

STAFFORDSHIRE

Left: On 14 July 1974 ex NCB locomotive *Wimblebury*, Austerity HE 3839 of 1956, shunts stock in the old colliery sidings at Foxfield, including the aged MW 0–6–0ST 1317 of 1895 and a diesel shunter, formerly petrol, by MR No 2262 of 1923.

Diesels normally work the traffic at Littleton Colliery, Huntington, Staffs, with Austerity HC 1752 of 1943 NCB No 7 retained as spare. The photographs on these two pages were taken on Saturday, 16 November 1974, on one of the colliery open days, which are held occasionally, with No 7 being steamed to convey visitors along the steeply-graded branch to Penkridge in brake vans which are retained for use on trains which cannot be continuously braked.

Right: Cabside of No 7. Authority to work over BR tracks in the exchange sidings is shown by the registration plate dated 1957, but locomotives are subject to periodic examination by BR inspectors. Overhead electrification warning plates are carried because the exchange sidings are situated alongside the BR Wolverhampton to Stafford line.

Left: No 7 shunts in the colliery yard. Operations over this cramped system are controlled by radio-telephone from a central point at the colliery with the shunters carrying pocket transmitters/receivers.

Right: Firebox cleaning at the end of the day.

Midlands stronghold—Cadley Hill Colliery, Swadlincote, near Burton-on-Trent, where normal roles in the seventies have been reversed with a diesel being held as standby to the steam locomotives, which are maintained in tip-top condition, both mechanically and externally.

Above: Cadley Hill No 1 snorts past 0–6–0ST *Empress* (WB No 3061 of 1954) standing in light steam outside the shed on 25 June 1974.

Below: Builder's plate fixed to bunker side on *Empress*.

Left: Youngster at work on 19 August 1974. Manufactured in 1962 by HE, No 3851 Austerity *Cadley Hill No 1* was the last but two steam locomotive delivered to the NCB, being originally built with an underfeed stoker which was subsequently removed. The aesthetically conscious Cadley Hill fitters removed the cone-shaped chimney and applied a standard one from a withdrawn engine.

SOUTH MIDLANDS

Deep in the Principality Austerity WB 2758 of 1944 rumbles towards the level crossing over the A48 Swansea–Carmarthen Road at Pontardulais with coal brought down the Dulais valley from Graig Merthyr Colliery on 24 May 1973.

WEST WALES

Earlier the same day, at nearby Brynlliw Colliery, Groves-end, 1916 built P 1426, an 0–6–0ST, is busy at the landsale yard.

Another view of P 1426 at Brynlliw.

Back at Pontardulais WB 2758 (left) is seen alongside her younger sister *Norma* HE 3770 built 1952.

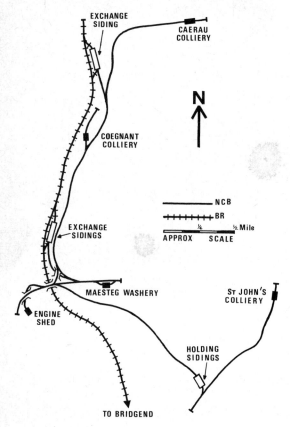

EXCHANGE
SIDING

CAERAU
COLLIERY

N

COEGNANT
COLLIERY

——————— NCB
+++++++ BR
¼   ½ Mile
APPROX   SCALE

EXCHANGE
SIDINGS

MAESTEG WASHERY

St JOHN'S
COLLIERY

ENGINE
SHED

HOLDING
SIDINGS

TO BRIDGEND

Left: Diagram of railways in the Maesteg area, where on the NCB lines the ubiquitous Hunslet-designed Austerity class had charge of the work during the final years of steam operation. The section of track from the headshunt by the engine shed to the holding sidings, near to the junction for St John's Colliery, was formerly part of the ex-Port Talbot Railway's Port Talbot–Pontyrhyl line, and was taken over by the NCB from BR in 1964, although prior to this the colliery owners had held running powers over this route. Stabled at the engine shed is privately preserved ex-GWR 0–6–0PT No 9642 which makes occasional steamings over parts of the system.

Below: Linda HE No 3781 of 1952 reverses a load for Maesteg washery out of the colliery yard at Coegnant on 24 May 1973.

Right: On the same day *Linda* is captured by the camera soon after passing Coegnant Colliery, propelling empties for Caerau. In the foreground is the connecting line to the exchange sidings.

Below: Linda again on 24 May 1973, this time blasting a coal train up the incline towards the washery, while on the right *Pamela* HE 3840 of 1956, is also at work. The leading wagon is about to pass the signalbox which controls movements at the washery.

Above: Pamela trundles along a section of the ex-PTR as she nears Maesteg with coal from St John's Colliery, which lies in the hills above the engine, on 7 October 1971.

Right: Pamela appears to have used up all her resources in bringing the train (seen above) over the difficult road from St John's as *Linda* comes to the aid of her sister in pushing the train, after reversal in the headshunt, towards the washery.

Below: On 24 May 1973 *Linda* clings to the hillside, bringing more coal for washing from Coegnant Colliery. In the foreground is the BR line and the Llynfi River.

EAST WALES

Right: Situated near the head of the deep Rhondda Fach valley is Mardy Colliery, Maerdy, where on 8 October 1971 a Peckett OQ class 0–6–0ST, P 2150 of 1954, the largest type built by this company for the UK, is at work.

Left: Viewed from the other side of the valley the previous day, No 2150 pulls slowly away from the exchange sidings with empties for the colliery.

Below: Looking a little forlorn as she rests at the back of Mardy shed on 23 May 1973 is ex-GWR 0–6–0PT No 9792, a few months before removal for scrapping. Nine years after being withdrawn from service by the Western Region her British Railways emblem can still be discerned on the tank side.

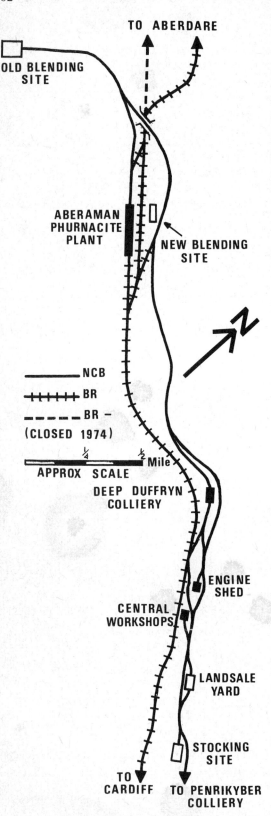

TO ABERDARE

OLD BLENDING
SITE

ABERAMAN
PHURNACITE
PLANT

NEW BLENDING
SITE

———————— NCB

+++++++ BR

– – – – – – BR –
(CLOSED 1974)

APPROX SCALE

¼   ½ Mile

DEEP DUFFRYN
COLLIERY

ENGINE
SHED

CENTRAL
WORKSHOPS

LANDSALE
YARD

STOCKING
SITE

TO
CARDIFF

TO PENRIKYBER
COLLIERY

Left: Diagram of railways in the Mountain Ash area.

Above: With a cataclysm of sound on Sunday, 27 October 1974 (when the photographs on these two pages were taken), 0–6–0ST No 1 (HC 1885 of 1955) struggles into the old blending site at Aberaman. This section was closed shortly afterwards when the new blending site became fully operational.

Top right: 0–6–0PT No 7754, built by NB in 1930 for the GWR and withdrawn from main-line service at the end of 1958, waits with its load opposite the phurnacite plant on track which once formed part of the GWR Pontypool–Neath line. Passenger services over this route were withdrawn by BR from 15 June 1964.

Right: No 7754 plods steadily away from the stocking site heading north. The line south had just been re-opened in order to link Penrikyber Colliery, Penrhiw-ceiber, with the system. This colliery is on the opposite side of the River Cynon to the NCB line, coal being carried across by an overhead conveyor to the loading bunker.

Right: Maximum effort from *Sir John* AE No 1680 0–6–0ST, built 1914, leaving the sidings at Mountain Ash on 5 October 1971. On the right is No 7754, behind which can be seen the engine shed and beyond the winding gear of Abergorki Colliery closed 1967.

Left: Austerity NCB No 8 (RSH No 7139 of 1944, rebuilt 1961 by HE No 3880), pushes coal from Deep Duffryn Colliery to an exchange siding now situated within the closed platforms of the former Taff Vale Railway Mountain Ash station, on 22 May 1973.

Left: Moment of respite for engines and men outside Mountain Ash shed on 26 May 1973. The engines are No 1 and nearest the camera *Sir Gomer*, P 1859 of 1932.

Above: Shunting at Merthyr Vale Colliery, Aberfan, on 6 October 1971 is NCB 0–6–0ST No 6 (P No 2061 of 1945).

Right: On the same day No 1 AB 0–6–0T, 2340 of 1953 pushes coal from Merthyr Vale Colliery, visible on the extreme right of the photograph, towards the exchange sidings. On the left is the River Taff and in the foreground the BR Cardiff–Merthyr line.

Above: 0–4–0ST hybrid locomotive, in front, built from parts of EV 2, P 1465, and P 1524, being handled by fellow 0–4–0ST P 1676 of 1925 on a rain-swept morning at Tymawr Colliery, Pontypridd on 23 May 1973.

Below: Painted with very noticeable warning bands AB 0–4–0ST, works No 1619 of 1919 *The Blaenavon Toto No 6*, hustles wagons out of the sidings at Blaenavon Coal Preparation Plant on 22 May 1973.

Right: On the same day *Toto No 6* fights for adhesion whilst charging towards the screens at Blaenavon.

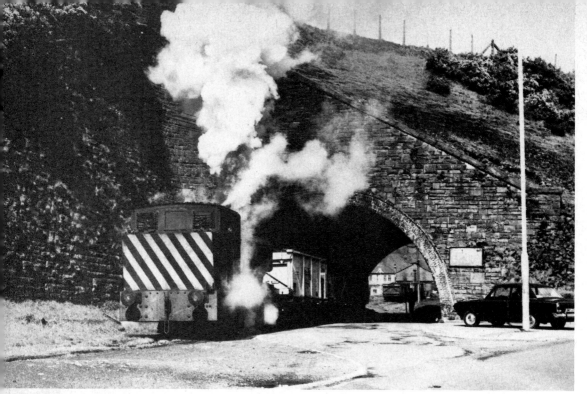

Engines from the locomotive shed at Talywain Landsale Yard, Abersychan, Nr Pontypool, worked the traffic over the long difficult line to Blaenserchan Colliery until 1970, when the output was diverted to Hafodyrynys Colliery, and by the time of my visit on 21 May 1973 (when the photographs on these two pages were taken) the day's work could be accomplished in about two hours. The depot closed in June 1974.

Above: Islwyn, a 1952-built 0–6–0ST AB, works No 2332, shunts outside the yard and under the tunnel which carries overhead the BR Pontypool to Blaenavon line.

Right: The valley resounds to the exhaust of *Islwyn* climbing laboriously away from the yard with empties for the exchange sidings. Notice the safety valves set at an angle.

Below: Islwyn being coaled by what is often the only means available.

Above: An unidentified Austerity at Hafodyrynys Colliery, Pontypool, on 4 October 1971.

Top right: Kilmersdon Colliery, Radstock, Somerset, administered by the East Wales area in recent years, was connected by a short branch and a self-acting incline to the former GWR Bristol–Frome line. On 13 October 1970 a solitary wagon is about to descend the incline as 1929-built 0–4–0ST P No 1788 stands by. The rope to the winding house can be seen trailing on the left.

Right: P1788 at Kilmersdon on 12 October 1971. The colliery, the last in the once extensive Somerset coal-field, closed in September 1973, following which the NCB handed the locomotive on permanent loan to the Somerset & Dorset Railway Museum Trust, based on the West Somerset Railway. In August 1974 the engine was named *Kilmersdon* in honour of the many years spent at the mine.

Our tour of the coalfields concludes with the photographs on these two pages at Snowdown Colliery, Nonington, in the Kent coalfield, Britain's smallest and newest, from which no coal was wound until 1913.

KENT

Left: St Dunstan eases hungry wagons past the screens on 16 September 1974.

Bottom left: On the following day as *St Dunstan* and HE Austerity No 3825 of 1954 stand by the weighbridge, and BR DE Class 33 No 33046 waits for the road into the exchange sidings, SR Class 432 750V dc 4VEP unit No 7870 hurries by, forming the 8.10 Victoria to Dover stopping train. The overhead wires are used by electric locomotives visiting the sidings, which normally take current from the third rail on the main line.

Above: Nameplate carried by 0–6–0ST AE No 2004 built 1927. The engine was named after the 25th Archbishop of Canterbury *St Dunstan* by the original owners, Pearson & Dorman Long Ltd, who named other locomotives after early archbishops and local saints.

Below: HE 3825 battles against the gradient with more newly-hewn coal on 16 September 1974.

Acknowledgements

In completing a book of this kind I am very mindful of the liberal assistance which I have received from many people, to all of whom I owe my thanks. It would be impossible to mention everyone by name, but I must place on record my sincere thanks to the officials of the NCB who have generously given me much help, both at headquarters in Hobart House and locally, and for their ready permission to use some of the photographs for publication. My gratitude is also due to the drivers and shunters I have met on my travels, and who, by their hospitality, have enhanced the enjoyment of the visits considerably.

I am indebted to many of my colleagues in the Industrial Railway Society and the Industrial Locomotive Society, and the pocket books published by the former have been of immense value.

My special thanks are due to Fred Atkinson and Mrs Mavis Phillips for their help and advice on photography matters, Derek Waugh who drew the maps, and Mrs Marlene McPherson who kindly typed the manuscript.